ALCHEMY AND THE GOLDEN WATER

STEVEN ALEX SCHOOL

DEDICATION

I dedicate this book to those who wish to study alchemy and who seek the philosophers stone. I am writing a series of books with each one revealing a little more information. Illuminating the celestial ruby which I myself have found. This is my second book upon this subject, the first book is available on amazon and kindle, it is titled alchemy and the green lion. There are many people studying alchemy these days, many of them are confused or simply on the wrong path, before you judge my words, simply understand that I have in fact discovered the truth of the stone with study, research, and hands on experimentation. If you follow my works you will undoubtedly discover the stone for yourself. Clear your mind and prepare to follow in my footsteps as I point the way to how I discovered the stone.

CONTENTS

ACKNOWLEDGMENTS

I would like to acknowledge amazon for helping me to publish books, I would like to thank Hermes Trismegistus for inventing our art, Michael Sendivogius for his works that truly are a guiding light, and my dog Belle, the only one who sticks with me through thick and thin.
THIS BOOK IS MEANT ONLY FOR ADVANCED ALCHEMISTS.

1 VEGETABLE REALM

When I first began my research into Alchemy, I didn't know what a menstruum was, or any of the types thereof. I began my studies researching and my first subject was lemon balm. I later moved into the animal realm which was a great learning experience, but I discovered there is nothing needed for me in that particular realm. Although I did learn some valuable lessons there about calcination, and centric salt.

In the vegetable realm of plants, I discovered three basic menstruums and they are alcohol, vinegar, and distilled water.

I myself chose absolut alcohol.
It does a wonderful job of extracting the essence of lemon balm from the dried leaves.

My basic procedure which was my own interpretation of how it should be done, and having no official guidance from other persons, I used a 2000 ml borosilicate flask with a rubber stopper. I ground the lemon balm as fine as I could get it in a mortar and pestle, I placed it in the flask and covered it with absolute. The alcohol turned of a nice golden color, I noticed if I left it soak longer, the color would change to green gold as the liquid acted upon the chlorophyll of the plant matter. Using a Buchner funnel and another flask for separation, I then placed the leftover plant matter in a suitable fire proof dish and lit it. The flammable alcohol burned the plant material and then I placed it in an oven for continued baking. When I was satisfied that the ashes had been baked to a light gray, almost white and having no blackness remaining, I then allowed the dish to cool, placed the ashes in the flask with the extract.

Heating the substance and stirring it to dissolve the centric salt, I filtered this solution.

I now evaporated this substance to the consistency of honey. And this was my tincture. My rendition of the primum ens mellisa.

We will now be skipping over the animal realm because it was a work of composing a menstruum by the ancients which simply is not needed in this day and age. In this modern age we have menstruums which are superior to those that were used in the past. If you are under the belief that our work requires a vile substance, open your mind and think again. If we cannot learn, we cannot grow intellectually. The old animal method slows and impedes the work. Therefore we will be advancing directly towards the mineral kingdom in the following pages. My works are not meant to be long winded and confusing as most alchemical writings are. My work is meant to steer you in the right direction in your search for that which is hidden. I will give you some ideas and principles to help light the way. Prepare to follow in my footsteps.

THE TEMPLE OF AMUN RA AT KARNAK.

The temple was a place of processing the golden water of the alchemists. They often referred to this water as the first matter of the stone, it is actually the second. The true first matter is the prima materia which you should have found by now on your own if you read my first book alchemy and the green lion. The menstruum is simply a tool that is used to work upon the prima. After the final extraction the original prima is discarded. Which leaves us with our water, our first matter according to the ancients. It is very much like basic spagyrics, only slightly different. Picture a Buchner funnel on top of a flask, now ponder the words of Hermes Trismegistus in his emerald tablet. "that which is above is like that which is below, to do the miracle of one thing". The temple of Amun at Karnak, this city of temples took about 2000 years to build and was completed some 4000 or so years ago. It was dedicated to Mut, Khonsu, and Amun. This impressive and awe inspiring structure is so large, that you could fit many large modern buildings inside of it. The interesting truth about the temple of amun at karnak, is that it was a place where alchemy was performed. My focus of study upon the temple of amun is that of deciphering the hieroglyphs which are recorded there. What most people do not realize is that the entire ancient process of creating the lapis philosophorum is in fact recorded in those glyphs, which is exactly where, and how I discovered it. Also revealed in those symbols is the true prima materia, and the menstruum of the ancients, what it was, and its creation. As I said before, I do not use the original menstruum, I have discovered through study and research, a modern solution which is superior to the original.

Ramesses 1 ordered the carving of some of the glyphs, some of these were called cartouches, and if you look at them with an alchemical point of view, you may notice as I have, some of them represent the solar disc, (the alchemical sun). the green lion symbol is also found as well as bowls, trevets, folded cloths, baskets, baskets covered with folded cloth and set upon a trevet indicating a filter, and rams headed sphinxes. Alchemists have said they're secret menstruum is found in the belly of the ram, which is indicated by the sphinxes, which are also known as the guardians of science because the menstruum is the key which unlocks the stone. King sety 1 ordered glyphs carved which pay tribute to his father, many of these symbols are still visible today. There is a very deep alchemical meaning in the glyphs for those that have eyes to see. Look at the cloth covered basket set upon a trivet, compare it to a modern Buchner funnel set upon a flask. As above, so below, truth revealed. Its not difficult now to understand hermes in his emerald tablet where he says, that which is above, is like that which is below, to do the miracle of one thing. We will cover this principle more in process.

MERCURY DUPLEX.

When you get to the point where you have found the golden water, upon its separation of the waters from the waters, (which is achieved through coction), use a Buchner funnel with a coffee filter set upon a flask, these filters fit the funnel perfectly. Stack 3 filters thick.

Carefully pour all of your golden water through the filter, once you are finished, look at the colored substance in the filter, as well as the water, and ponder as above, so below. The moon is in the water, the sun above. Gabritius and beya. The two ingredients of the stone. Now continue extracting from your prima with fresh menstruum, many times until it will extract no more. Saving all of the sulphur principle above. Take your golden water, calcine it to ashes until it doesn't smoke any more. Dissolve it in hot distilled water and filter it. Repeat this process of calcine, dissolve, filter, calcine, until your salt is prepared as clear as liquid crystal with no more blackness remaining. When your salt is thus prepared, dissolve it in absolute alcohol, this is your duplex rectified.

NOTE-
the colored sulfur principle can be white or citrine, it is the body of the stone.

I will divulge another great secret about that powder in my next book, which will be titled "The menstruum of the philosophers".

PROCESS

Fill a crock pot with clean sand, Place your sulfur principle in a 250 ml, borosilicate flask. Bury your philosophical "egg" halfway in the sand so that you can look into it. The top of the egg is to remain open at all times. (the ancient adepts stressed that your egg must be hermetically sealed so as to confuse you. In truth, the egg remains open. Now add a little of your rectified duplex to the egg, this is imitating the rain cycle. Allow the duplex to evaporate, then add a little more, continue repeating this process until you have used all of the duplex and the stone is coagulated. The crock pot should be set on low the entire time.

Your philosophers stone is now finished and completely multiplied.

This is the true process of the ancients, for confecting the stone, the only difference is we have modern equipment and electricity. I decoded the process myself from the ancient glyphs at karnak.

In the words of Michael Sendivogius,
SIMPLICITY IS THE SEAL OF TRUTH.

MULTIPLIERS.

As we discussed earlier, the sulphur principle is already the body of the stone. There are two multipliers, choose only one of them according as you wish to work white or red. The first multiplier is simplex rectified. It is the multiplier of the white stone.
The second is duplex rectified, it is the multiplier of the red stone. We already discussed how to prepare the duplex.

To prepare the simplex rectified. Take your original golden water, from the first filtration and uncalcined. Gently evaporate it to dryness on low heat like that of warm sunlight. Upon reaching dryness the green lion will be exposed. Place the green lion in a distillation train, you will distill this, and collect only the white fume which coagulates in the cold. This is the secret white water which has power to calcine the sun. it is also the multiplier of the white stone. (don't forget to extract centric salt from the ashes).

Sometimes nature has done some of our work for us. After you filter the golden water, subject it to coction on low heat for a few days. In summer just leave it out in the hot sunlight. If the water separates and becomes clear with a substance floating in it, filter it. Rejoice, nature has simplified and shortened your work. If the powder is white, use this as the body of your white stone and multiply it. If it is citrine colored, yellow, orange, or red, use it as the body of your red stone and multiply it accordingly. To test the red powder, throw some of it on a very hot iron plate. It should subtilize into a fine scarlet powder which is completely unaffected by fire.

The red and white sulphurs are the mysterious red and white powders of the ancient alchemists. Two stone jars containing these substances were actually found in Nicholas Flamel's basement.
I have given you plenty of guidance if you have discovered the prima, and chosen a menstruum.

8

THE GREEN LION.

The green lion is the ore of Hermes, it is the entrance to the philosophers garden, and according to the ancient sages it is also the starting point of the work, however it must be prepared by art, therefore the obtainment of this lion is the true starting point of our process. I pondered this lion for a few years, considering the possibilities, and experimenting to test my theories, before I finally found it. The writings of the ancients seemed to lead in all the wrong directions. In truth, this lion is a waxy green substance. In scientific terms, it has been identified and given a modern name, although the modern scientists probably do not know the alchemical value of this particular substance, they simply know that it is an element, and that it has been given a name. the lion has also been called the tree of life, and the tree of knowledge. In the writings I have seen recommendations to look under the vegetable roots in springtime, that basically in a fertile meadow filled with vibrant colorful flowers and green grasses in spring, that one would carefully cut and peel back the mat, dig down two feet, collecting that soil, then take two pots, drill holes in the bottom of one, and set it on top of the other pot, cover the holes with cloth to create a filter and a receptacle underneath, now place this earth into the perforated pot, and pour boiling rainwater, or boiling distilled water, through this earth, that it will dissolve something that nature has put in this dirt and pass it through the filter. Then when all has cooled, collect the filtered water and gently evaporate it to dryness on low heat. It is said that a salt will be obtained, which is the sulfur of the earth. I have always been intrigued by this but have never yet attempted this to see what is produced. I have also read of another method which is a winter application, it is said that if one explores the snowy areas of the wilderness where gold has been found, that you may come across a gelatinous mass, sitting atop the snow, which has no obvious explanation of what it is, or why it is there. The ancients called it gur, which is slang for green lion, it rises through the pores of the earth, it is congealed by the frost of snow, upon thaw it is set free again to rise into the heavens. I have seen this substance myself, one year before I began studying alchemy, I remember wondering about this strange gelatinous mass, at the time I had no idea what it was. If you can catch it, great, however there is another substance in nature which contains this substance, nature has been at work to place it there under lock and key, closely guarded and protected from prying eyes, it is ignored, and disregarded by most, considered worthless trash, except to the alchemist who has eyes to see it for what it really is. Within itself it contains all that we seek, the androgynous matter containing the two metallic natures, and the astral spirit, Which is the green lion.

There is said to be another method which is performed usually in the fall, it is to be done at night with a black light. The practitioners of this method will choose A large unimproved open area of raw land, the African plain would be a perfect place for this, or a desert, mountainous regions where gold is found would be another good place. Basically when the season has changed and plant growth is withering as we head towards winter, it means the celestial life force containing salt is leaving the earth, headed back up into the solar system in its continual cycle of recharging itself with the universal energy so that it can return next spring to once again bring health and vigor to our world. As it rises through the soil during the day by the heat of the sun, at night it cools and rests upon the surface of the soil, a black light is said to reveal this luminous white salt just laying there on top of the soil, and that it can be scooped up and collected. It can be further purified as white as snow by subliming it in and aludel.

This substance, many have called, AZOTH.

General equipment utilized in the great work. These are the things I use, listing them is designed to save you money by avoiding the purchase of unnecessary items which may be expensive. Cast iron mortar and pestle, this is the very first tool that you will need, mine is from the civil war, I came across it on ebay, apparently soldiers in the field used these to make black powder for the cannons, however it is perfect for alchemy. Also a clear glass mortar and pestle will be required. You will need a 250 ml borosilicate retort, you will also need to construct a homemade aludel, I have not found anyone who makes these, it is also wise to invest in a metal retort for higher heat distillations, mine is copper. I have a copper alembic which works well, but a borosilicate alembic is superior, however here is what you should look for in a good alembic made of glass, the unit should come with both a blind head, and a distillation head. The manufacturer should have heated the distillation head and depressed it slightly, thus creating a rain gutter inside the distillation head so that everything which rises, can drip only into the condensing arm, and not back down into the curcurbit (base), of the alembic. Also when your alembic is disassembled, you should be able to easily fit your entire hand inside the curcurbit, the distillation head, and the blind head. To achieve this you will have to get this alembic custom made, this is the Cadillac of distillation vessels for the great work, this will greatly speed the great work. If you cannot reach inside this vessel, you are in for some frustration, because you will see a glorious substance form inside this glass, but you will not be able to collect it, look but don't touch so to speak. And we want to be able to reach in, and scrape out this beautiful salt.

I save glass pickle jars, these are inexpensive and perfectly suited for our work, no need to buy an expensive flask with a ground stopper just for extractions, when in use cover the tops of these jars only with saran wrap, if you seal them tight, expanding gases with crack the glass and ruin your work.

An assortment of funnels, glass, plastic, and ceramic Buchner funnels as well. Clean spice jars with labels are good for storing separated and purified substances. Coffee filters are perfect for use with the Buchner funnel. Cheap electric hotplates, and a portable plumbers torch will generally cover most of our applications, a gas burner will sometimes be needed, I use the side burner on my propane grill.

Rubber stoppers, an assortment of these will be necessary, it can be difficult to find them with the correct sized holes in them, so what I do is use an assortment of flat spade drill bits with a cordless drill and just drill my own holes.

A cheap electric crock pot filled with sand is used for incubating our eggs, mine has four heat settings, warm, low, medium, and high. I find that the warm setting is perfect, like the warmth of a hatching chicken.

You can find an assortment of eggs at online auctions, get the 250 ml size with the extra long neck, it needs to be a round bottom flask with a rubber stopper, the ideal neck length is six inches. You can usually find these for about five dollars each, I recommend that you get about a dozen of them to start with, or more. A serious alchemist may easily have sixty or seventy experiments going at one time, thus expediting the learning curve, and speeding the accrual of knowledge.

To simplify multiplication, it is imbibition. This means that we have two substances, one of which is a solid and the other a liquid, with this it is simple to figure out which one is the body of the stone, and which one is the multiplier. I have purchased a small clear glass bottle with a built in eye dropper for performing the imbibitions, with this inexpensive tool I can accurately gauge the precise amount of liquid that I want to dispense.

I recommend having a borosilicate flask as a receptacle for your retort, and a separate flask with a flat bottom and a ground glass stopper, for storing the astral spirit which comes over the helm.

Sol is reduced into mercury by mercury. The viscous water made thick, dissolve and coagulate, three turns of this alchemical wheel. Nature does the work, let us examine this information, the correct substance must first be obtained, and then it must be processed and refined and purified until it leaves us with two things which I will describe for you now.

The first substance is sol, it is the body of the stone, it is a white foliated earth, the sages liked to turn things backwards or upside down to help confuse everyone, so they at times mismatched the color, texture, or sex of each element, resulting in much confusion and difficulty for anyone trying to understand this portion of the great work. They said the moon was in the water, they sometimes said the water is red or yellow, even at times calling it the white wife, but in fact the water is crystal clear like tears.

Once you obtain our correct substance, and process it into these two, the white foliated earth, and the crystal clear water, they are joined together in the egg, in a warm sand bath. At basically body temperature, but not to exceed 110 degrees Fahrenheit. The process does not take years or several months as the sages claimed, it actually goes very fast.

You place one part of the white foliated earth (sol), into the egg, then add three parts of the water, if you add more water than this it is fine, it will take longer to congeal, but it will produce a higher multiplication of the stone. Another and perhaps better way to gauge the ratio is thus, place the salt in your egg, then add enough water that completely absorbs the salt and dissolves it into itself, so that you now have only water. The egg at this point needs to be no more than one third filled at the maximum, we need room for further multiplications of the stone.

In a short time, measured in weeks, (less than a month), this water will congeal into a fixed stone, once it is completely dry and fixed, it is considered white but it actually looks more like fine silver, only transparent.

At this point in time, as the stone is completely fixed and dry, you may slowly begin to turn up the heat by degrees, do not be tempted to turn it up too quickly, the gradual, small increases of heat, will incite the color changes to occur, they are not as the philosophers wrote. You will first see citrinatas, a yellowing of the matter. It will become like yellow glass, then as you continue to increase the heat very slowly by degrees, the matter will continue to darken to blood red, the celestial ruby, at this point in the work, we could actually darken it to the point that it looks like glittering black glass, however shine a bright light through it and you will see that it is translucent blood red. If this stone is ground to powder it will revert to a nice lemon yellow color, if it is re melted in the low heat of a candles fickle flame, it will turn as red as the darkest ruby you have ever seen, and upon cooling it will once again be a beautiful, translucent red stone. This is the stone of the first order, it is suitable to be used to create the medicine for man, however for metals we must spin this alchemical wheel two more times, repeat the imbibition, the dissolving and coagulating again, exactly as you did the first time, reduce the heat, add the water until your substance is completely dissolved, each repetition of this multiplies the stone to the next order, each multiplication increases its power by tenfold. We can multiply this stone more if we want, most alchemists chose anywhere from the stone of the third order, to the stone of the seventh order. The higher stones can still be used for medicine, however they must be further diluted accordingly.

As each multiplication augments its power tenfold, each dose of elixir must also be diluted an extra tenfold for each level.

Keep in mind in alchemy work that the sages wrote in metaphors, so do not use common mercury, gold or silver in your work, any pathy that says to use common mercury or cinnabar is false. It also has been proven time and time again, that those who believe the stone to be created from urine are mislead. Hennig brandt performed multiple experiments upon urine in search of the philosophers stone, he never found it because it is not there, he discovered phosphorus instead.

Also, many times I have denounced the star regulus of antimony path, many people follow this path and it leads them to nothing, the truth of it is that this path, as it is written and taught is incorrect, simply because people always seem to forget that the sages wrote in metaphors, in code if you will, so you must understand that the ingredients listed for the work, is false. The incorrect ingredients are listed in the recipe, if you will find the way, you must learn the correct substances to fill in the blanks, you must do this part on your own, no one freely divulges that knowledge as to do so would incur a great and possibly unforgivable sin, however take heart in knowing that it really is not difficult to find that information on your own if you simply apply the required effort and refuse to quit or give up. I found the truth myself, and if I can do it, so can you. Once the correct elements are placed in the egg, an endothermic reaction occurs, the substance begins to produce heat from within itself, this is a type of nuclear fusion, causing the generation of its own heat, and these two substances fusing together into a new substance, which did not exist before. This baby is a toy of our own, nature does not produce this stone in the wild. It takes nature a thousand years or more to complete the work of producing metals within the bowels of the earth, by attempting to replicate the process, the alchemist plans to do this above ground, and in less than one years time, even though the plan is to replicate natures process as closely as possible, never the less, the way of the alchemist is still slightly different than that of nature. Some changes must be made, some corners must be cut, if we are to perform in a few months of time, that which normally would have taken a thousand years or more.

There are also two basic wet paths to the stone, the ars longa or original long way, and the ars brevis, or abbreviation of the work. They are in fact the same path, however some alchemists accepted the way they were taught in the original method, and handed this down to their disciples, but some other alchemists performed additional experiments upon the matter since they had a thirst for knowledge, so they experimented further into the mysteries of the hermetic art, and what they discovered was that some of the steps in the magnum opus, were actually not needed and could be eliminated from the process, thus shortening the work. I myself have found at least eleven unecessary procedures which I have now eliminated from my work, and I expect that there may be more shortcuts, therefore if one of my writings seems to slightly contradict another, this may be a clue for you to look deeper into this portion of the work, and potentially discover an abbreviation.

It has been written that there have been master alchemists in the past such as maria the prophetess who were said to be able to complete the great work from start to finish in as little as three days, however those secrets were kept very well hidden. The only way to discover them would be through much experimentation, and then only maybe at best, however it is not impossible or unattainable, I myself have not reached nearly that level. My alchemy works are part of a highly specialized subject, they are not very long, more like booklets. But I intend to delve more directly and fully into the subject of the stone than most authors who fill volumes with useless filler and never even find the stone. My cover pictures are my actual work. The proof is in the pudding. The picture on the cover of sol and luna the hermetical wedding, is the true conjunction of sun and moon, the picture of the red stone on the cover of alchemy survival guide is my genuine red stone, naysayers and false prophets can dispute my work with negative reviews, doubting the truth of my work, but look at the cover of my book, alchemy and the green lion for the final outcome of this art. As I said, the proof is in the pudding. I wish someone would have written a series of books like this for me when I was coming up in this art, for I had many questions, I searched far and wide to discover that there simply was no one, who had the answers. Perhaps all the real alchemists had left us centuries ago, this left me in the position of understanding that if I wanted the knowledge of this art, I was going to have to get up off of the couch, and actually discover the truth myself. It isn't that no one was willing to help me, it was that no one had the answers to give, I had to blaze a trail and cut through the veil of secrecy and hidden knowledge on my own because there is simply no one else at my level in this art. The stone is a lost art, and the artisans have long since left the building, they have moved on to some other plane many years ago, and all they left us were some vague and confusing writings which really did not seem to convey any real knowledge, such as the works of geber, which is where the word gibberish stems from, people trying to decipher the writings of geber equals useless gibberish, and with good reason, the alchemists wrote their notes in secret codes which only they could understand, it was never meant to teach anyone else this art, they were just recording their findings for their own later reference since we may tend to become forgetful later in life. However once the final mastery of the basic stone is achieved and has been performed to completion a few times, it is like riding a bicycle, you never forget. There are many finer points in the work however, such as the generation of all types of gem stones and medicines, not to mention the creation of new species. The Egyptians were said to have created the cat in the pyramids, as well as some other creatures. There are writings which suggests that they may have made old people young again, and even resurrected the dead.

Our finished product in this art is known as the great red wax, when heated it will melt and flow like wax with no loss of substance, and no smoke or fume. It has power to tinge red hot metal, which is the final test of our substance.

Other books by Steven School.

Sol and luna the hermetical wedding.
Alchemy and the green lion.
Alchemy and the peacocks tail.
Alchemy survival guide.
The secret recipe book, kitchen tool box.
Grandmas delicious recipes.
Trophy wife.
Casino survival guide, breaking the bank.
How to make money.
Wilderness survival tips.
Booze survival guide, a coffee table book.
Karate secrets revealed, knowledge of the masters.

We will reserve a few pages here for your study notes. Remember, to do the great work you need two things, a liquid and a solid, the prima materia and a suitable menstruum, each phase of the work revolves around a liquid and a solid.

Good luck in your journey,
S.A.S. 2013

The Magnum Opus DVD
www.createspace.com/381734

Alchemy And The Athanor DVD
www.createspace.com/388897

www.ingramcontent.com/pod-product-compliance
Lightning Source LLC
Chambersburg PA
CBHW051305170526
45165CB00004B/1865